I0493744

PREFACE

From the various ways in which one can express about any life experience, poetry is one of them.

The poems, which you(the reader) are about to witness, have being

arranged in a chronological order(as mentioned in index),on basis of their formation as well as on the basis of my perception getting clearer on varied fields of reality.

ACKNOWLEDGEMENT

I thank everyone, who directly or indirectly have inspired me in writing the poems.

AUTHOR

Kevin Shenoy

INDEX

1) AN EXTRAORDINARY YESTERDAY OF 'I'
   (1st January)

2) THE RISING TIDE
   (10th February)

3) CONSTRUCTING POWER
   (24th February)

4) AN EVER HAPPENING FACTOR
   (11th March)

5) ABSTRACT IS THE NEW REALITY
   (18th March)

6) RE-ENTER IN THE MATHEMATICAL THOUGHT
   (26th March)

7) FRIENDSHIP OF THE BEST FORMS
   (5th August)

8) FUNCTIONS IN THE FLOW
   (10th September)

9) BEING HAPPY FOR NO REASON

   (24th October)

10) IN ABSOLUTE DEVOTION TO EVERYTHING

   (13th December)

AN EXTRAORDINARY YESTERDAY OF 'I'

Yesterday did 'I' realize;
What means and what demeans.

Yesterday did 'I' realize;
How to see the deceive and the truth in deep.

Yesterday did 'I' realize;
How hard it is to transform and how easy it is to deform.

Yesterday did 'I' realize;
For how lonely 'I' is after all, for all that is done in truth is tough to penetrate a

surface of strings hardbound.

Yesterday did 'I' realize finally;
What means Gnostic Atheism and what means Gnostic Theism!

THE RISING TIDE

The tenebrosity which is to be deviance;
Since from darkness comes the radiance.

The melancholic agony which haunts
in this spontaneous experience;
Of being stuck in, but another illusionary
spectre in grievance.

The origin is extraordinarily fun-
damental of this standstill;
That causes the oddity in wholeness of the frills.

Since the inaugurator of this disastrous sulk;
Is none other than one, own-self responsible in
bulk!

So exploring self is the constant at-
tempt to rejuvenate and to affirm;
That the luminosity is stronger for any maudlin
to diminish or to curb.

The overwhelming power of this will;
Dismantles all forms of the integrated woe frag-
ments, which drill.

The vigour of waves of truth, converts
the numerous revolts in the mind;
Into the sparkling sects of energy, which com-
bine in to from a source which is to shine.

Thus, the calmness attained;
 Is called infinity in elate!

CONSTRUCTING POWER

The something which is left out even after an halt.

The ingredient that strengthens in infinity along with the exact accurate, equal resonant energy.

The point at which most of the phenomena can be experienced.

The shadow that covers the original, from emitting the actual real light so as to deteriorate and desperate the false might.

The parallel piercing pieces at par with the poise !

The noise of the voice which enters in a vacuum and there it lies.

The altitude at which, impossibility is of trivial existence.

All these lead to the emergence of 'power', pro-
gressing in build.

AN EVER HAPPENING FACTOR

The phase of transition in reminisce sparkles;
The enclosure of which linearly tackles.
Since the snap of the wind flows in rumble;
And still the tenet is on the halt, which is better
if in crumbles!

ABSTRACT IS THE
NEW REALITY

To confront the interminable, necessi-
tates an undiluted mettlesome valour;
Which in turn devises an
aura of lofty certitude.
When overweening arrogance
is subject to plummet in uproar,
Squalls of calmness swells in the amplitude and
the altitude.

When a metaphysical milieu fur-
ther envelops in magnitude,
Nowt can quieten this mathematical urge
Thus, extremes of every-
thing inhabit in infinitude;
Guileless and unworldly in gurge.

To plunge and baffle the resistance in submerge,
Causes the involute deviances to torment.
Cyclical distortions culminate and diverge,

Kevin Shenoy

When the theoretical frameworks flash in
torrent.

Abstract is unalloyed and more real;
Since no other reality can overhear!

RE-ENTER IN THE MATHEMATICAL THOUGHT

Infinity is the search;
In any Real or Abstract submerge.

To tend to the Abstract, hence to the Pure;
Aligns one, collinear to the turbulence of the Storm!

'For the Storm' which is Real
in destruction and tone;
There exists, a noteworthy definition of it's own.

This implies Reality is always Abstract;
And the Abstract is rather more Real, for all the Exact!

Thus, the Abstract can also be termed as Mathematics;
Which firmly holds in account, it's Statistics.

Since in this unparalleled 'true' Abstract,

all the Real as well as the Abstract phenomena are subject to Theories and, are
attempted to explain on the basis of Proofs!
Henceforth 'mostly' true and to this, the Abstract reacts," To contradictions, I bid the fore-
most formal adieu!'

So if to say, "Mathemat-
ics is a Storm!", is not wrong;
But it is definitely different in origination, based
on Axioms!

FRIENDSHIP OF THE BEST FORMS

Why should emphasize on friendship be laid, just, on the fifth of August?

Why does a certain day need to express the vitality of something that is infinite, when the concept is about 'best friends'?

Why does such a day need to prevail, when pre-dominant is the foundation of formation?

Wouldn't this day seem trifling, if it is to exhibit the degree of association of something immense and immeasurable?

Why does it have to be a one day phenomenon, rather an eternal phenomena?

And so, why does the term friendship need to be isolated to a specific day of every year, and instead be a everyday happening in liveliness?

FUNCTIONS IN THE FLOW

If any of the trigonometric functions tend to 'happiness', then none of them will ever be able to case it. Since, in a periodic annihilation will they all be always entrapped in.

When, exponentially does the 'happiness' function (y=exp(x)), then there exists a need for 'sorrow' to behave as the logarithmic function(y=ln(x)), such that
the former paces with a laid fountion, and the latter comes in rather slower approach, in chase.

Now with this continuation, if the y=x 'curve' is reasoned in treason, to be a fair judge to both the curves, when it itself is a 'line' unknown to the twists and curls. it depicts a weird philosophy that ventilates:"Joy and Sorrow are 'equidistant'!", which by all means is a non-integrable dour way to stage!

If the function is to 'ceil' or 'floor' the quantity(y-=ceil(x),y=floor(x) respectively), then the discontinuity seems reasonable, as the jumps play off the play of rapid descend of peace.

And, if haste isn't the grave entailment, then the modulus(y=|x|) of the variable is the best plausible way to express the 'joy', since 'gloom' never exists in it's interest!

BEING HAPPY FOR NO REASON

A psychological imbalance as it may seem, for a being to joyous, without any reason to deem. But in terms of life, if exuberance showers it's light, only when associated to a(some) particular 'thought(s)', then there lies a potential in the way of the mind to break apart the 'self', whenever only the intellect applies itself to lead! To be enslaved by intellect, defines the detachment from joy, to which the stature of dolour possesses an expansion to petrify!

To be in an endless bliss for everything; substantially paves way to life, and also saves one from the actual mental illness of being in a constant state of 'thinking' about:

"If only the world was good enough to understand me;

No grief would have ever dared to absorb in me!"

To be addicted to this remark, deprives one from the reality; thus being ignorant to life itself!

The life, which is left out to explore about.
The life, which is of immense possi-
bility to focus on to be devoted into.
The life, which desires the complete intensity of
one, in whichever fields one energizes in.

If one is exultant forever, unaffected by anything that comes as an obstacle, well then:

Should then, a reason be either held responsible
for or irresponsible for one's glee?

Should then, either, people be blamed for, or
appreciated for the depreciation of one's happiness?

Should then, a reason even exist, for a being to
suffer about?

 If not, then one understands that the 'fault' lies
within, when gloom envelops one; and not in
anyone or anything to account for.

Every piece of life is bestowed with a soul to live
the 'life'. Sense lies in this journey, if it's full of
radiance. But the 'common' sense in the current
time lies in, if sorrow is seen in everything!

IN ABSOLUTE DEVOTION
TO EVERYTHING

In a boundless cosmos, 'motive' predominantly
stimulates to 'differentiate' everything that
comes in it's perception.

When the armour of 'ego' expands around;
How will the fluidity in activity surround?

When emotions lean on to be biased;
How will the thought process commute to that
in the same dimension, in interest?

The frequency at which ecstasy resonates
is directly proportional to the vibrancy
of embedding everything in totality;
The happening of this in form of 'moments', or
as an unceasing 'life' connotes, how does one
comprehend the entire existence which pre-
eminently is incompressible in acceptability.

www.ingramcontent.com/pod-product-compliance
Lightning Source LLC
Chambersburg PA
CBHW040916180526
45159CB00010BA/3090